Chicken Processing Plant Fires
Hamlet, North Carolina
and
North Little Rock, Arkansas

Investigated by: Jack Yates

This is Report 057 of the Major Fires Investigation Project conducted by TriData Corporation under contract EMW-90-C-3338 to the United States Fire Administration, Federal Emergency Management Agency.

Department of Homeland Security
United States Fire Administration
National Fire Data Center

U.S. Fire Administration Fire Investigations Program

The U.S. Fire Administration develops reports on selected major fires throughout the country. The fires usually involve multiple deaths or a large loss of property. But the primary criterion for deciding to do a report is whether it will result in significant "lessons learned." In some cases these lessons bring to light new knowledge about fire--the effect of building construction or contents, human behavior in fire, etc. In other cases, the lessons are not new but are serious enough to highlight once again, with yet another fire tragedy report. In some cases, special reports are developed to discuss events, drills, or new technologies which are of interest to the fire service.

The reports are sent to fire magazines and are distributed at National and Regional fire meetings. The International Association of Fire Chiefs assists the USFA in disseminating the findings throughout the fire service. On a continuing basis the reports are available on request from the USFA; announcements of their availability are published widely in fire journals and newsletters.

This body of work provides detailed information on the nature of the fire problem for policymakers who must decide on allocations of resources between fire and other pressing problems, and within the fire service to improve codes and code enforcement, training, public fire education, building technology, and other related areas.

The Fire Administration, which has no regulatory authority, sends an experienced fire investigator into a community after a major incident only after having conferred with the local fire authorities to insure that the assistance and presence of the USFA would be supportive and would in no way interfere with any review of the incident they are themselves conducting. The intent is not to arrive during the event or even immediately after, but rather after the dust settles, so that a complete and objective review of all the important aspects of the incident can be made. Local authorities review the USFA's report while it is in draft. The USFA investigator or team is available to local authorities should they wish to request technical assistance for their own investigation.

This report and its recommendations were developed by USFA staff and by TriData Corporation, Arlington, Virginia, its staff and consultants, who are under contract to assist the USFA in carrying out the Fire Reports Program.

The USFA greatly appreciates the cooperation and information received from Hamlet, North Carolina Fire Chief David Fuller and Captain David Knight, State Fire and Rescue Division Deputy Commissioner Timothy L. Bradley and Deputy Director Ray Eastman and Special Agent David H. Campbell of the State Bureau of Investigation. Thanks also go to officials of Tyson Foods, Inc., Shannon Weathers, Mike Edmunds and Mike McAlister.

For additional copies of this report write to the U.S. Fire Administration, 16825 South Seton Avenue, Emmitsburg, Maryland 21727. The report is available on the USFA Web site at http://www.usfa.dhs.gov/

U.S. Fire Administration

Mission Statement

As an entity of the Federal Emergency Management Agency (FEMA), the mission of the U.S. Fire Administration (USFA) is to reduce life and economic losses due to fire and related emergencies, through leadership, advocacy, coordination, and support. We serve the Nation independently, in coordination with other Federal agencies, and in partnership with fire protection and emergency service communities. With a commitment to excellence, we provide public education, training, technology, and data initiatives.

 FEMA

TABLE OF CONTENTS

Chicken Processing Plant Fires
Hamlet, North Carolina
(September 3, 1991)
North Little Rock, Arkansas
(June 7, 1991)

Local Contacts: Hamlet, North Carolina Fire
Chief David Fuller
Captain David Knight
Hamlet Fire Department
302 Champlain Street
P.O. Box 1229
Hamlet, North Carolina 28345

Deputy Director Ray Eastman
Special Agent David H. Campbell
Arson Division
North Carolina State Bureau of Investigation
P.O. Box 29500
Raleigh, North Carolina 27626

Timothy L. Bradley - Deputy Commissioner
Fire and Rescue Division
North Carolina Department of Insurance
P.O. Box 26387
Raleigh, North Carolina 27611

North Little Rock, Arkansas Fire
Shannon Weathers - Fire Safety Coordinator Loss
Control
Mike Edmunds - Corporate Safety Director
Mike McAlister - Plant Manager – Prospect Farms
Tyson Foods, Inc.
P.O. Box 2020
Springdale, Arkansas 72765-2020

OVERVIEW

The morning work shift of employees at the Imperial Foods Processing Plant in Hamlet, North Carolina, had just begun when a fire occurred, at approximately 8:15 a.m., on September 3, 1991. The rapid spread of heavy smoke throughout the structure ultimately resulted in 25 fatalities and 54 people being injured in varying degrees. Of the people who died, 18 were women and seven were men.

SUMMARY OF KEY ISSUES

Issues	Comments
Casualties	Twenty-five fatalities and 54 people injured in varying degrees.
Building Structure	Poultry processing plant of 30,000 square feet with open work areas, sealed concrete slab floor, ceramic tile walls, and ceilings of formica-type finish. Interior kept cool.
Origin and Cause	The conveyor to a cooker had hydraulic line repaired which burst when brought up to full pressure. Hydraulic fluid expelled at 800 to 1,500 psi, ignited by heating gas plumes of a cooking vat.
Fire Spread	Immediate and very rapid spread of heavy black smoke throughout the building.
Evacuation	Fireball and rapid spread of smoke caused disorderly evacuation attempts. Several exit doors locked, drove employees to seek refuge in cooler or seek other exits. Rapid build-up of toxic gases killed personnel attempting to escape.
Detection and Alarm	Plant Operations Manager found phone line already inoperable, ran to vehicle and drove to fire station.
Response	Rapid response by combination department and people from immediate community including medical personnel who ran from nearby hospital. Search and rescue delayed by heavy smoke and heat. Ample mutual aid from neighboring communities, including helicopter transport of victims to regional medical facilities.
Code Enforcement	During the 11-year operation of the plant, no inspection conducted by North Carolina Occupational Safety and Health Administration.
Critical Incident Stress	Debriefings provided through Pee Dee Council of Government. Many fire service personnel involved in incident knew or were related to the victims

A similar type fire occurred at a chicken processing plant in North Little Rock, Arkansas, on June 7, 1991, but with no fatalities or injuries. Following the description of the Hamlet fire below, the North Little Rock fire is summarized along with the factors in the different outcomes of these two fires.

THE BUILDING

Imperial Foods occupied a one-story brick and metal structure that over the years had been used for various food product operations. Reportedly, the previous operation had been in dairy products. As such, the interior work areas had walls, ceilings, and floors conducive for that type of operation. This meant that these three surface areas were of materials that could be washed down. The floor was a sealed concrete slab, the walls were ceramic tile, and the ceilings were a formica-type finish. The total square footage was approximately 30,000. For the layout of the plant see Appendix A.

Imperial Foods operations did not include the slaughter of poultry. Rather, poultry parts were shipped to the plant, which prepared and cooked the chicken. The cooked chicken would then be distributed to various markets for use in restaurants.

The plant had a total employment of approximately 200 people, with a normal shift having around 90 employees. Preparation of the poultry products included trimming, marinating, cutting, and mixing. The prepared meat would then be cooked, quick-frozen, packed, and prepared for shipping. Storage areas varied from large drive-in coolers to quick-freezing units.

The plant layout allowed easy movement of products from one area to another by electrical forklift pallet movers. The entryways between the various preparation areas were for the most part open while some entrances had a curtain of plastic strips to assist in holding refrigerated air in the rooms. The freezers and coolers had standard refrigeration doors.

The preparation areas were for the most part cooled or refrigerated in order to prevent food spoilage. Accordingly, door openings were designed in a manner to seal in the structure, with door seals similar to those on a refrigerator. This was necessary to assist in maintaining a constant temperature in work areas.

Day to day the contents inside the building did not represent a major fuel load problem. The only combustible products were items such as paraffin-coated shipping boxes and wood pallets. Therefore, the probability of having an extensive fire was considered remote.

The bulk of the food processing operations was performed in the south three-fourths of the complex. The north one-fourth was predominantly for storage and loading. The main operations areas by virtue of their cooled, open rooms did present a problem in that there were no smoke or heat barriers between work areas. This meant that in the event of any type of fire, there would be nothing to impede the travel of heat and smoke. Furthermore, the predominance of hard, smooth surfaces meant there was little available material to absorb heat and smoke.

There were exterior personnel doors throughout the structure. These included the main entrance on the east side; the southeast loading and trash compacting dock; doors from the break room and the equipment room to the outside; and a door from the packing room which led to the north one-fourth of the building complex. However, the locations of some of these exits and their sizes would in all probability have excluded them from being considered appropriate as part of an evacuation plan.

THE FIRE

The area identified on the Floor Plan in Appendix A as the processing room is the room where the fire incident occurred. This area is centrally located within the building complex. Any incident occurring in this area could adversely affect much of the building operations and personnel.

Poultry products that had already gone through the various marinating and mixing procedures were taken by conveyor to a cooking vat in the processing room, which contained soybean oil. The oil was maintained by a thermostat control at a constant temperature of 375 degrees Fahrenheit plus or minus 15 degrees Fahrenheit.

A maintenance worker who survived the fire indicated that the hydraulic line that drove the conveyor had developed a leak. The hydraulic line was turned off and drained of fluid. Then the maintenance worker disconnected the leaking line and replaced it with a factory prepared line.

The factory prepared line, however, was found to be too long and would have dragged on the floor, possibly causing people working in the area to trip. So the maintenance worker reportedly asked for and gained permission to cut the factory prepared hydraulic line to an appropriate length, replaced the end connector with their own connector, and put the line back in-place. This line has been described as a 3/4-inch flex line rated to carry 3,000 psi. Information from plant personnel indicated normal pressure was kept at approximately 800 psi, but it would at times fluctuate as high as 1,200 to 1,500 psi.

The hydraulic line was brought back to operating pressure. Shortly afterward it separated at the repaired connector point. The connector was some four to six feet above floor level with hydraulic fluid being expelled at a pressure of 800 to 1,500 psi. It obviously began to splatter off the concrete floor. Droplets were bouncing back onto the gas heating plumbs for the cooking vat, which turned them into vapor. The vapors then were going directly into the flame. The vapors had a much lower flashpoint than the liquid hydraulic fluid and therefore rapidly ignited.

In sum, the pressurization of the hydraulic fluid combined with the heat was causing an atomizing of the fuel which in all probability caused an immediate fireball in and around the failed hydraulic line and the heating plumbs.

The ignition of the fuel caused an immediate and very rapid spreading of heavy black smoke throughout the building. Seven workers were trapped between the area of origin and any escapable routes.

Measurement of the system during the investigation after the fire indicated 50 to 55 gallons of hydraulic fluid fueled the fire before electrical failure shut the system down. (Investigators stated that if the hydraulic system was fully charged and its reservoirs filled to capacity it would have held 110 gallons of 32 weight ISO hydraulic fluid.)

In addition to the hydraulic fluid, the fire reached a natural gas regulator that in turn failed and caused an induction of natural gas to the fire increasing the intensity and buildup of toxic gases.

The fires in both this incident and the North Little Rock incident were centered around the cooking vat areas and expanded outward from there. In both incidents, the vats ultimately did ignite in latter stages of the fire, but in the initial stages the vats did not ignite. The vats in both locations have a hood-mounted system over them with built-in CO_2 heads. But after considerable burning with secondary falldown, the oil in both vats did eventually ignite and burn.

Witness reports indicate much of the plant was enveloped in under two minutes. Workers throughout the plant found their visibility eliminated and oxygen quickly consumed. Hydrocarbon-charged smoke, particularly as heavy as this, is extremely debilitating to the human body and can disable a person with one or two breaths. This was confirmed as autopsies conducted on all of the fatalities found that virtually all died of smoke inhalation as opposed to direct flame injury.

Survivors indicate there was no real organization in the plant's evacuation, and this was confirmed by the locations of the bodies. Several employees in the central part of the structure moved to the trash compactor/loading dock area near the southeast corner of the building. It was here they found one of the personnel doors to the outside locked. A trailer was backed into the loading dock cutting off all exiting through this area. One woman became trapped between the compactor seal and the building wall while trying to squeeze through an opening. A number of remaining people in this area went into a large cooler adjacent to the loading dock, but failed to pull the sealed door shut thus allowing smoke infiltration into the cooler. The cooler had the largest single fatality count area with 12 deceased people being removed from this room along with five injured people.

The second largest fatality area were the seven trapped in the processing room between the fire and any escape route. Three additional bodies were found in the trim room area, one of whom was a route salesman who had been filling food machines in the break room. The exterior personnel door in the break room was the other door locked from the outside.

The people who died in this tragic fire were as follows:

Name	Age	City
Josephine Barrington	63	Hamlet
Peggy Anderson	50	Hamlet
Mary Lillian Wall	50	Rockingham
Philip R. Dawkins	49	Rockingham
Minnie Mae Thompson	46	Hamlet
Janice Marie Wall Lynch	43	Hamlet
Elizabeth Ann Bellamy	42	Bennettsville, SC
Cynthia S. Wall	41	Rockingham
Josie M. Coulter	40	Rockingham
Bertha Jarrell	40	Rockingham
John Robert Gagnon	39	Hamlet
Rose Marie Gibson Peele	39	Bennettsville, SC
Mary Alice Arnold Quick	38	Hamlet
Fred Barrington, Jr.	37	Rockingham
Martha A. Ratliff	36	Hamlet
Gail V. Campbell	33	Hamlet
Rosie Ann Chambers	31	Ellerbe
Michael Morrison	31	Hamlet
Rose Lynette Wilkins	30	Laurel Hill
Brenda Gail Kelly	28	Rockingham
David Michael Albright	24	Hamlet
Margaret Banks	24	Laurinburg
Donald Bruce Rich	24	Ellerbe
Jeffrey Antonia Webb	24	Hamlet
Cynthia Marie Ratliff	20	Hamlet

FIRE SUPPRESSION AND EMERGENCY MEDICAL SERVICES (EMS)

Upon discovery of the fire, the operations manager of the plant attempted to call the alarm to the fire department, but found that phone lines were already inoperable. (Imperial Foods was not equipped with pull-station alarms, nor does the town have 9-1-1.) He then ran to this vehicle parked outside and drove some three to five blocks to the fire station.

The initial equipment left the station at 8:24 a.m., and was on the scene three minutes later. Fire Chief David Fuller indicated the first smoke he observed was grayish-yellow in color. He stated that Hamlet has two paid firefighters on-duty at all times with 28 volunteers. Of the 28 volunteers, 22 responded to the scene. (See Appendix B for Fire Department Incident Report.) He also stated there is a county mutual aid agreement and that Captain Calvin White immediately called for the Rockingham Fire Department to stand in at the station. Lieutenant David Knight indicated that upon their arrival on the scene they immediately encountered three dead-on-arrivals and 15 to 18 casualties. Their first actions were to administer first-aid and attempt victim rescue. Once they had backup companies on the scene, the fire was attacked. Extremely heavy volumes of smoke prevented them from pinpointing the seat of the fire in the early stages. AFFF foam was used to extinguish the vats which eventually caught fire. Upon seeing the magnitude of the incident, additional mutual aid assistance was called in, including the East Rockingham, Cordova, and North Side Fire Departments.

In addition, two EMS units were initially brought in by volunteers with a third unit added later. Shortly afterwards, a call was made to the County Sheriff's Office to call all available EMS units to the scene. These consisted of two from Rockingham, three from the county, one from Cordova, one from Ellerbe, and one from Hoffman. Also, there were helicopters from Winston-Salem, Chapel Hill, Duke, and Charlotte which took patients from the hospital in Hamlet to the various burn units. The helicopters did not operate from the scene.

Chief Fuller stated that the City of Hamlet did not have its own inspectors and relied on one of the county's three inspectors. The county has an inspector for building codes, another for electrical and another for plumbing. These inspectors are primarily for new construction or remodeling. Hamlet construction codes reference the Southern Building Code. Chief Fuller stated the local code requires "periodic" inspections but do not specify a schedule or frequency.

The original building at Imperial Foods was built in the early 1900's. Today no one appears to know what codes existed when the plant was first built. Chief Fuller indicated there had been several fires in the plant over the years, some before Imperial Foods took over the facility. Imperial was operating the plant in 1983 when one of the previous fires occurred after which a CO_2 system and hood over the cooker was installed. Subsequently, they were required to install a CO_2 system by the county inspector.

Firefighters immediately began a search and rescue operation but were met by considerable heat and fire coming from the processing area. They had to withdraw and reposition to initiate their attack on the fire through the equipment room which was next to the processing room. The fire was brought under control at approximately 10:00 a.m.

Search and rescue efforts continued during the fire suppression with injured people and fatalities being located from the first entry at approximately 8:45 a.m., with the final victim being located shortly after 12:00 noon. Concern for the integrity of the roof structure prevented earlier discovery of victims in the processing room area.

Treatment of casualties was being carried out during the entire incident until all were removed from the fireground. Word of the incident's severity spread through the community quickly, and virtually everyone involved with medical care in the area responded to the plant site. The Hamlet Hospital is approximately six blocks from Imperial Foods.

Chief Fuller was asked to evaluate the handling of the incident with reference to fire suppression, rescue, and EMS to which he indicated he felt there were more than adequate numbers of personnel and equipment given the layout of the incident site. As it was, he stated there were minor problems of some EMS equipment running over charged hoselines. There was some problem later in the fire suppression with air for the self-contained breathing apparatus (SCBA) supply because tanks were being used to assist injured victims, as well as supplying fire personnel. Chief Fuller stated the entire incident centered around one problem – lack of enforcement of existing codes.

DISASTER PLAN

Hamlet does have a disaster plan in-place which coincides with the county plan. This fire occurred so rapidly and was so serious that all of the resources planned for were immediately brought to the scene. Under the plan the mayor is in charge of media relations and this became an enormous task, which was handled well even though much greater demands were being made upon them than had ever been planned for.

STRESS MANAGEMENT

The Hamlet Fire Department personnel suffered severe stress and emotional reactions because the community was small and the firefighters knew many of the victims. A critical incident stress debriefing was arranged through the Pee Dee Council of Government (Region H). Five counselors were brought in and 50 to 60 people attended the counseling sessions.

CODE ENFORCEMENT HISTORY

Much discussion has taken place about the lack of inspections conducted by the North Carolina Department of Occupational Safety and Health Administration (OSHA) at the Imperial Food operations. In fact, during the 11-year operation of this plant, North Carolina OSHA had never inspected the facility.

NORTH LITTLE ROCK, ARKANSAS FIRE

During the course of the investigation of the Hamlet, North Carolina, fire, information was received that another company had experienced a similar fire at a plant in Arkansas. Further inquiries revealed that indeed a similar fire had occurred at a Tyson Foods, Inc., facility in North Little Rock, Arkansas, on June 7, 1991, but with dramatically different results.

Tyson Foods, Inc., is the largest producer of poultry products worldwide. Corporate officials strongly believe that their safety program is what made the difference between their fire and the Hamlet, North Carolina, fire. This company has over the years enacted many proactive fire safety programs.

Their operation and plant type in North Little Rock is similar in product production but larger in size than the Hamlet plant. The fire that occurred on June 7, 1991, was in the same plant area as the Hamlet fire in that it broke out in the hydraulic system of their cooker (also made by Stein and Associates, as was the one in Hamlet).

Unlike the Hamlet fire though, the hydraulic failure occurred within fixed plumbing. A flange type nut over time had the threads stripped as a result of vibration and when the threads failed, hydraulic fluid was expelled at approximately 800 psi. As with the Hamlet fire, the fluid was immediately in an atomization state. This was occurring within the gas heating plume areas and fire ignited. The resulting fire created a heavy black smoke and, as seen in Hamlet, virtually coated everything it spread to.

The Tyson plant, however, had in-place numerous safety factors that averted disaster. They enforced life safety codes to include not only plant design but practiced emergency drills.

Tyson Foods has a fire safety director who has implemented evacuation programs throughout the company's entire operations. These programs involved both hourly personnel and management staff in safety committees. They have formed fire brigades and have a program called the Incipient Fire

Force, which involves all personnel and has a common goal to educate and train all employees in loss prevention and to take proper action should an emergency occur.

On June 7, there were 115 people working at the North Little Rock plant with some 12 to 14 people in the packaging area above the production room where the fire occurred. The plant design was such that the minimal number of people needed to operate the cookers were the only ones in the actual ignition area. The cookers were in rooms with 2-hour fire rated walls and ceilings, and the cookers were fed by conveyors through small openings. When the worker on the cooker that ignited discovered the fire, he first reached for an extinguisher but immediately realized it was spreading too fast and sounded the alarm.

Within three minutes, everyone was out of the plant and supervisors immediately identified all employees by name to make certain all were accounted for. No injuries of any kind occurred. Upon fire equipment arriving at the scene, fire brigade members, wearing hazardous materials protective clothing and SCBA equipment, met the firefighters and guided them through the plant to the seat of the fire.

The initial response came from Station 4 of the North Little Rock Fire Department. Backup came from the North Little Rock central station and also from the Little Rock Fire Department.

They were on the scene for approximately 9-1/2 hours. There were a total of 23 fire service personnel who responded with a total of six engines, one piece of aerial equipment, and three other types of vehicles.

As with the Hamlet fire, heavy, black smoke quickly permeated the entire facility. The firewalls surrounding the cookers no doubt gave the people evacuating more lead time – this was part of their pre-fire planning in that the cookers were designed to be isolated as much as possible from the remainder of the plant. In addition, Tyson allows absolutely no combustible materials such as wood pallets or paraffin-coated cardboard boxes inside the cooker rooms.

These types of operations are viewed as a wet industry for the most part. Accordingly, much of the facilities are not sprinklered. Tyson's safety personnel did not feel that sprinklers would have contributed to the prevention of loss of life due to the nature of the hydraulic-fluid-fueled fire. They do, however, have sprinkler protection in all areas that are non-wet operations.

Damages to the structure amounted to approximately 8 million dollars and the plant was down for 13 weeks. The additional loss in production, wages, cleanup, etc., was approximately 4 million dollars making the total loss approximately 12 million dollars. But upon getting back to production, the remodeling of the plant eliminated certain inefficiencies and implemented numerous safety features beyond what they already had. The plant currently has 215 employees, which is slightly less than the work force at the time of the fire.

FIRE PROTECTION EQUIPMENT AND SAFETY PROGRAMS

The plant now has shut off valves designed for each cooker. These valves have four functions in that they are calibrated to the hydraulic fluid velocity or flow of what each cooker needs or uses. Should there be 1) a sudden free flow of fluid; 2) a drop in line pressure; or 3) an electrical failure, the valve will shut the hydraulics down. It is also tied into the CO_2 system. In addition, they have mandated that if a system is installed by an outside manufacturer, then training must come from the manufacturer on maintenance of the item.

In addition to the shut off valves, Tyson had remote hydraulic shut down switches installed at strategic locations throughout the plant and next to pull alarms. Any one of these emergency switches being activated immediately shuts down ALL hydraulics in the plant.

Emergency lights were in-place above all exits. At the time this report was prepared, Tyson officials were considering adding a second emergency light lower to the floor as an extra assist should there be a sudden induction of wet, heavy smoke, as experienced at the June 7 fire.

They already had negative air flow pressure systems for ammonia releases, which are activated by sensors. They indicate rapid heat rise sensors could be added to exhaust heat and/or smoke.

Tyson Foods requires that each of their facilities have a minimum of two fire drills a year; most of their plants do it on a quarterly or monthly basis. When a drill is conducted, production is affected and some food products may have to be discarded to meet United States Department of Agriculture (USDA) inspections. Even so, Tyson Foods makes this mandatory. They have a formal safety policy, and, in addition, each plant has a required Monthly Fire Inspection Checklist they must submit to corporate headquarters. (See Appendix D.) The checklist covers many areas and must be signed off by the Fire Brigade Chief and Facility Manager. This type of checking helps them to detect deficiencies before they develop into problems. This was recently demonstrated in one of their Texas plants when a monthly checklist noted a drop in water pressure to their sprinkler system. Upon further inspection, they found that the city had changed their water usage classification and dropped their flow and pressure. Without a monthly checklist, this might not have been detected for months or until an emergency occurred.

Tyson Foods has a daily inspection of CO_2 systems. Also, they use permits that must be issued before any operation such as cutting and welding takes place. They also have lock-out, tag-out procedures that, simply stated, means before any repaired system is put back to use it is tested. Had they discovered the failed flange nut at the North Little Rock plant before the fire, they would have replaced the failed area, brought it back to operating pressure, and made sure it would hold before turning gas back on to the heating plumbs.

Tyson Foods has incentive/reward programs to encourage all personnel to be on the lookout for added safety ideas. The incentive bonuses are based on paperwork turned in in a timely manner, site visitations/inspections, and their participation in documented safety training. Tyson's has installed within their break rooms suggestion boxes for any complaints or suggested improvements. These are checked daily and responded to by the management staff of that facility.

Tyson Foods has Crisis Management Manuals in-place at each of their facilities. Plant management is totally familiar with each phase of those manuals and they, in turn, disseminate the information to each employee. Tyson Foods officials feel their emphasis on life safety also leads to protection of property and continuity of operations.

As a matter of coincidence, the last thing stated by one of the Tyson safety officers was that the rules and safety codes to protect personnel and property are already written and in existence, but for them to be effective they must be enforced. This was one of the final statements of Chief Fuller at the Hamlet Fire Department as well.

LESSONS LEARNED

1. Life safety codes must be enforced.

Life safety codes cover a broad range of topics but the main goals to be achieved are to 1) plan building layouts/construction to reduce hazards and have available the proper number of exits; 2) provide detection and adequate suppression equipment where needed; and 3) train and educate personnel in loss prevention and the proper action to take in the event of emergencies. The blatant problem of having exit doors locked on a continuous basis is clearly one that must be addressed by enforcement officials. Enforcement is an essential as the code requirements themselves. Enforcement can be by State or local officials. And in some cases, as with Tyson Foods, industry itself takes its own initiatives in both code enforcement and proactive fire safety programs in its plants.

2. Cooking areas must be separately partitioned from other employee work areas.

Any time there is a food processing plant cooking operation, with moving parts and high pressure equipment, the risk factor is greatly increased that a fire will occur at some time. As such, it is imperative that the cooker operations be partitioned off from the remainder of the building, and workers, as much as possible. The rebuilt Tyson plant designed their cookers to be inside 2-hour fire-rated walls with openings for incoming and outgoing food. Safety doors were installed and the minimal number of needed employees was all that were inside the cooker room. In addition, absolutely no combustible products were allowed inside the room.

3. Building exits in wet type operations should have double emergency lighting, one positioned above the door and one low to the floor.

A fast developing, heaving smoke was present in both of these fires. The work areas are kept cool according to USDA regulations for food preservation, so the relative humidity is high. These are described as wet operations. When heat-charged smoke is injected into this cool, damp air, it banks down more quickly than normal. People are taught from an early age that in the event of fire they should get as close to the floor as possible to maximize safe evacuation. Heavy smoke such as was experienced in these incidents obscures the upper lights at the emergency exits. Additional emergency lights in protective cages near floor level would assist personnel in locating exits. Consideration might also be given to having strobe lights as part of the emergency lighting system.

4. High pressure equipment maintenance and repairs must be limited to factory trained personnel and specifications.

Operations such as these plants have extensive hydraulic systems. They operate at considerable pressures and, as in these two cases, are integral parts of cooking processes. Moving parts and high pressures will naturally increase the likelihood of failure. Maintenance must be a constant ongoing process. For maximum safety, maintenance personnel must be trained by the factory representatives. If any parts are replaced, they must conform to factory specifications or not be used. If a maintenance division alters any part of the high pressure system, their alterations must meet or exceed factory specifications.

5. High pressure equipment in probable incident areas should have built-in catastrophic shut down valves.

Inlying valves have been designed that are sensitive to multiple functions for high pressure hydraulic equipment. These valves are calibrated to the prescribed hydraulic fluid velocity or flow of the equipment needed. The valves will automatically shut down fluid flow if there is 1) a sudden free flow; 2) a drop in line pressure; or 3) an electrical failure. In addition, these valves can be linked to the CO_2 fire suppression systems.

6. **Negative air flow systems in these facilities could enhance safety by being modified to also accomplish smoke evacuation.**

Many plants similar to those in the Hamlet and North Little Rock fires have negative air flow systems in the event of an accidental ammonia release. These are activated by sensors and can purge the area of toxic fumes very quickly. If, in addition to the gas sensor, a rapid rise heat sensor was added, these systems could pull the heavy, wet smoke away from the lower levels in the event of a fire.

7. **State and Federal inspectors from various departments should be cross-trained.**

Much has been said about the lack of inspections done prior to the Imperial Foods tragedy. The State OSHA inspector force was small in numbers and simply could not cover all of the industry. Yet there were USDA inspectors frequently present because of their responsibilities over the food processing industry. While it may not be possible to teach USDA personnel all aspects of an OSHA inspector's responsibilities, certainly they could be encouraged to recognize major problems while carrying on their assigned duties and alert the State OSHA office or other appropriate authorities including plant management.

8. **Establish a "worry free" line of communications for industry employees.**

Although it has not been acknowledged firsthand and was told only through the media, reports have surfaced that workers inside the Hamlet Plant were afraid to say anything about safety conditions due to fear of being fired. In order to eliminate this type of possible problem, communications with plant management or regulatory authorities should be established. States may wish to establish agencies/systems such as a known agency/address to write to or an 800 phone number. The identity of the individual reporting deficiencies could then be protected.

9. **The number of OSHA safety inspectors must be increased.**

As with many governmental departments, OSHA has had their budgets cut and funds have been directed to other areas. The tragedy of the Hamlet fire demonstrates vividly why cuts should not take place within the area of code inspections and enforcement. In North Carolina (and in other States as well), the number of staff should be based on the number of inspectable properties and time it takes to meet inspection schedules.

10. **Emergency exit drills must be incorporated into industry policies.**

The posting of emergency routes and exits throughout a structure simply will not suffice. The actual practice of the routes and exits must be done. The exit drills include a system to number and identify employees in order to make certain that everyone has been evacuated. The drills should be conducted often enough that employees will be constantly aware of emergency procedures. The actual practice of drills paid off for Tysons Food as they had all 115 employees out of the building within three minutes and all were accounted for, through a system of employee identification.

Imperial Foods Plant Floor Plan

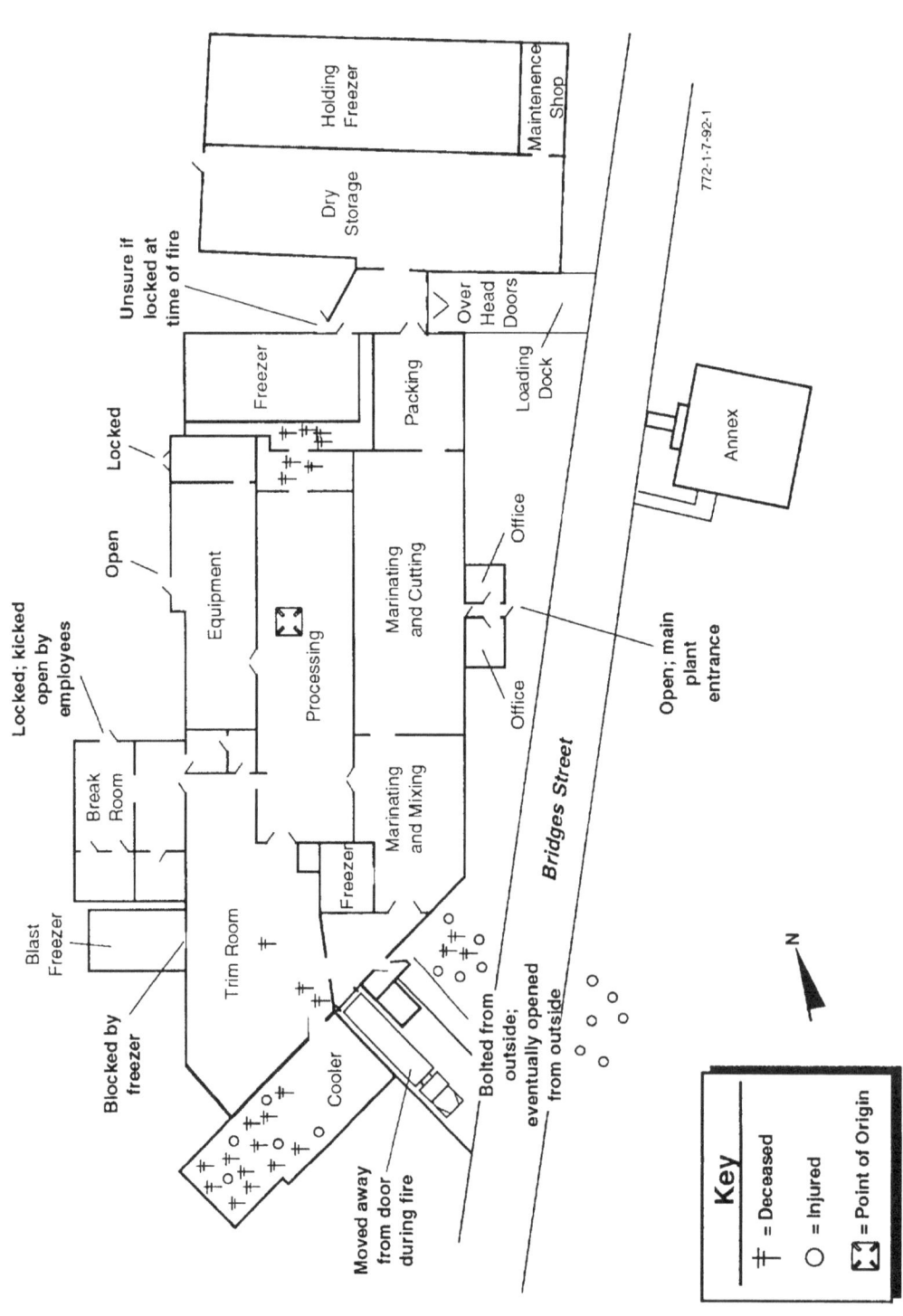

Holding Freezer

Maintenence Shop

Dry Storage

772-1-7-92-1

Over Head Doors

Loading Dock

Annex

Freezer

Packing

Unsure if locked at time of fire

Locked

Office

Open

Equipment

Marinating and Cutting

Open; main plant entrance

Locked; kicked open by employees

Processing

Office

Break Room

Marinating and Mixing

Bridges Street

Blast Freezer

Trim Room

Freezer

Blocked by freezer

Cooler

Bolted from outside; eventually opened from outside

Moved away from door during fire

N

Key

† = Deceased

○ = Injured

⊠ = Point of Origin

12

APPENDIX B

Hamlet, North Carolina, Fire Department Incident Report

NORTH CAROLINA INCIDENT REPORT	0 7 7 0 3	Hamlet Fire + Rescue					9 7 7 0 0 9 7 0 0 9 K

Mo.	Day	Yr	Day of Week	Alarm Time	Time Out	Arr. Time	Time In	Tot. Time Out	FIRE SERVICE RESPONSE
0 9	0 3	9 1	3	0 8 2 4	0 8 2 5	0 8 2 8	1 9 1 5	0 5 1	

INCIDENT ADDRESS OR LOCATION
Street: 40 Bridges Street — Rm. or Apt. — Personnel | 2 | 2
City: Hamlet — State: NC — Zip: 2 8 3 4 5 — 7, 11. — Engines | | 2

OCCUPANT NAME
Last, First: Imperial Food, Inc. — Phone: (919) 582-3552 — Mutual Aid (check one): 1 ✓ Received — Aerials | | 1

OWNER NAME
Last, First: Roe, Emmit — Phone: ( ) — 2 □ Given — Tankers | | |

OWNER ADDRESS
Street: 1155 Hammond Drive Suite 5230 — 3 □ Not Apply — Other Vehicles: Rescue 7,8,9 ST-1, ST-2, Ch21 | | 6
City: Atlanta, — State: Georgia — ZIP: 3 0 3 2 8 — Hazardous Materials Involved: 1 □ Yes 2 ✓ No

PLEASE PUT APPROPRIATE CODE NUMBER IN BOX FOR EACH CATEGORY

METHOD OF ALARM FROM PUBLIC	TYPE OF SITUATION FOUND		TYPE OF ACTION TAKEN	REQUIRES COMPLETION OF INJURY & FATALITY REPORT
1 Telephone	11 Structure fire	19 Fire/explosion not classified	1 Extinguishment	No. Incident-related injuries
2 Municipal alarm system	12 Any fire outside a structure where the material burning has a value	20 Overpressure rupture (no combustion)	2 Rescue [2] Primary	Fire Srv. 7 Other 55
3 Private alarm system	13 Vehicle fire	30 Rescue	3 Investigation	
4 Radio	14 Trees, brush, grass fire	32 EMS only	4 Remove hazard	No. Incident-related fatalities
5 Verbal	15 Refuse fire (material burning has no value)	40 Hazardous condition	5 Standby	Fire Srv. 0 Other 25
6 Home dialer	16 Explosion, no after-fire	50 Service call	6 Salvage	Is juvenile suspected in ignition?
7 Tie-line	17 Outside spill, leak with fire	60 Good intent call	7 Medical Aid [1] Secondary	1 □ YES 2 ✓ NO
8 Voice signal: Fire alarm system		71 False malicious	8 Fill in, move up	
9 Other [5]		73 False malfunction	9 Cancelled enroute	Is property abandoned or vacant?
		74 False unintentional	0 Water supply	1 □ Yes 2 ✓ No
		99 Other situation found [11]		

Fill in this section if "TYPE OF SITUATION FOUND" is 11, 12, 13, 16, 17, 19 ONLY (14, Optional)
(Refer to coding sheet)

		Fixed Property Use	
		MANUFACTURING	700
Ignition Factor: Accidental Fuel Spill	41	Area of Fire Origin: Processing	38	Equipment Involved in Ignition: Processing Equipment	70
Form of Heat of Ignition: Fuel Fired Object	10	Type of Material Ignited: Flam/comb liquid	20	Form of Material Ignited: Power transfer equip	60

If Heating Equipment Involved. Type of Fuel Used	1 Kerosene 4 Wood 7 Natural Gas 2 LPG 5 Coal 8 Gasoline 3 Electric 6 Oil 9 Other 0 Not Apply	[0]	PROPERTY DAMAGE CLASSIFICATIONS	Estimated Value
			1 $1-99	
			2 $100-999 [9] Value	Estimated Structural Damage
			3 $1,000-9,999	
CONDITION UPON ARRIVAL	MOBILE PROPERTY TYPE		4 $10,000-24,999 [9] Damaged	
1 Overheat	11 Automobile	20 Freight road transport 00 Not Apply	5 $25,000-49,999	
2 Smoldering	12 Bus	30 Rail transport	6 $50,000-149,999	
3 Open flame	13 All-terrain vehicle	40 Water transport	7 $150,000-499,999	Estimated Contents Damage
8 Out on arrival [3]	14 Motor home	50 Air transport	8 $500,000-999,999	
	15 Travel trailer	60 Heavy equipment	9 $1,000,000 or more	
	17 Mobile home	70 Special vehicles, containers	0 NO DOLLAR LOSS	
		99 Other mobile property types [00]		

If Mobile Property	Yr	Make	Model	St.	Lic. Number	Serial Number/VIN
If Equipment Involved in Ignition	Yr	Item	Make	Model		Serial Number

NO. OF STORIES	EXTENT OF DAMAGE	DETECTOR PERFORMANCE
1 Single Story	1 Confined to the object of origin	1 □ Present 2 □ Not Present
2 Two Stories	2 Confined to part of room or area of origin	
3 3 or 4	3 Confined to room of origin Flame [6]	If Present, Type of Closest Unit
4 5 or 6	4 Confined to fire-rated comp. of origin	1 □ Smoke 2 □ Heat
5 7 to 10	5 Confined to floor of origin	
6 11 to 20	6 Confined to structure of origin Smoke [6]	Power Supply
7 21 to 50	7 Extended beyond structure of origin	1 □ Battery 2 □ A/C
8 Over 50	9 No damage of this type Water [3]	1 In room of fire: operated
Below Grade		2 Not in room of fire: operated
...ding [1]	CONSTRUCTION TYPE	SPRINKLER PERFORMANCE	3 In room of fire: did not operate
...ht	1 Fire resistive	1 Equipment operated	4 Not in room of fire: did not operate
[1]	2 Noncombustible	2 Equipment in service, did not operate	5 In room: fire too small to operate
	3 Heavy timber	3 Equipment present: fire too small to operate	9 Not classified (Not Apply)
	4 Ordinary [4]	8 No equipment present	Fire Referred for Investigation to:
	5 Frame	9 Equipment not in service [8]	SBI 1 ✓ Yes 2 □ No
	0 Other		

...harge (name, position)	Member Making Report
P Fuller	Capt. Calvin White

13

APPENDIX C

Imperial Foods Plant Fire Photographs

Photo by Jack Yates

View of the south end, the southeast corner, and the front (east side) of the Imperial Foods building.

Appendix C (continued)

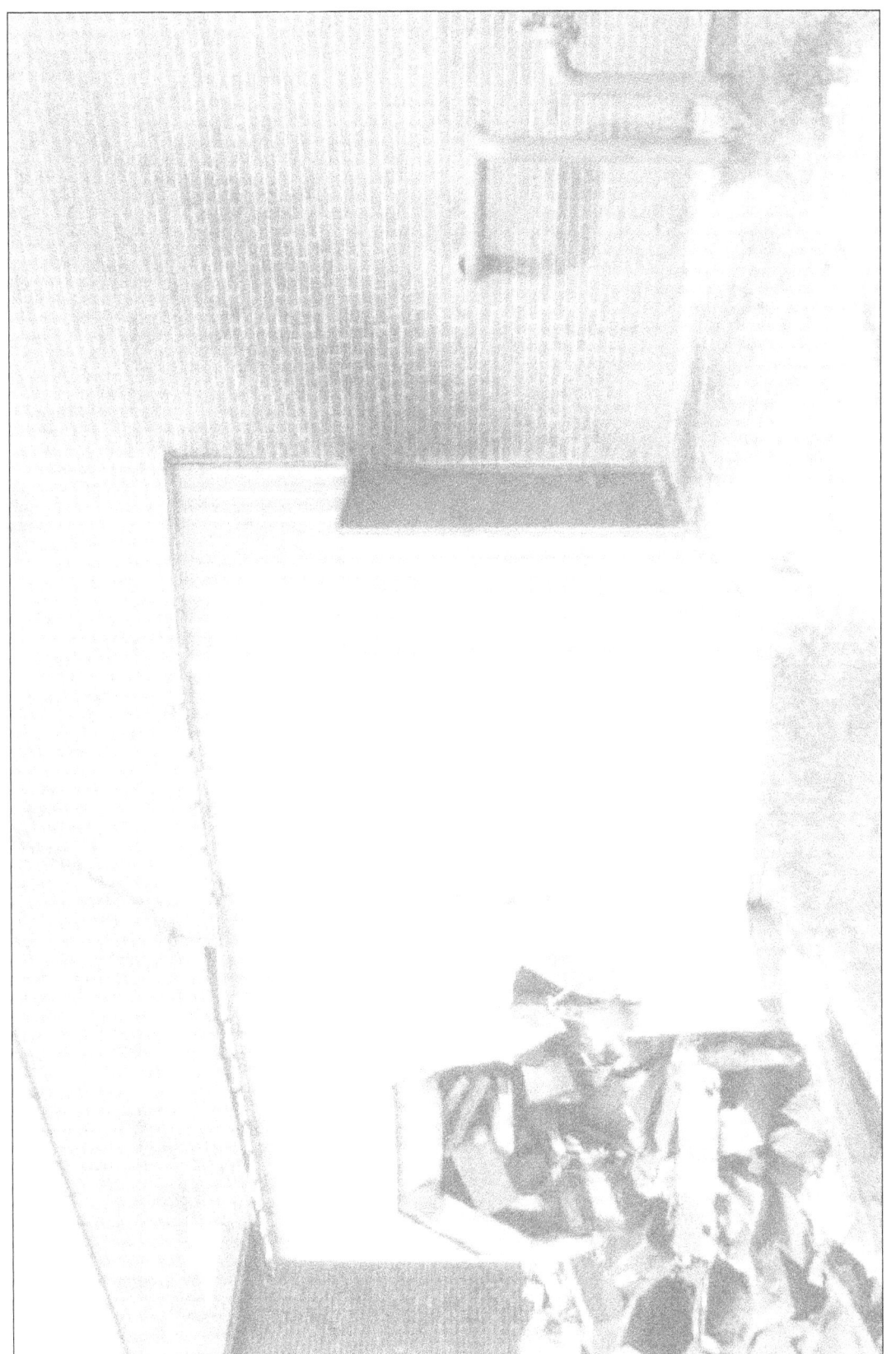

Photo by Jack Yates

The compactor area and one of the exit doors that was locked.

Appendix C (continued)

Photo by Jack Yates

The loading dock area and sealed doors next to the compactor. The pads around the doors are designed to form a seal when the trailer backs into this dock.

Appendix C (continued)

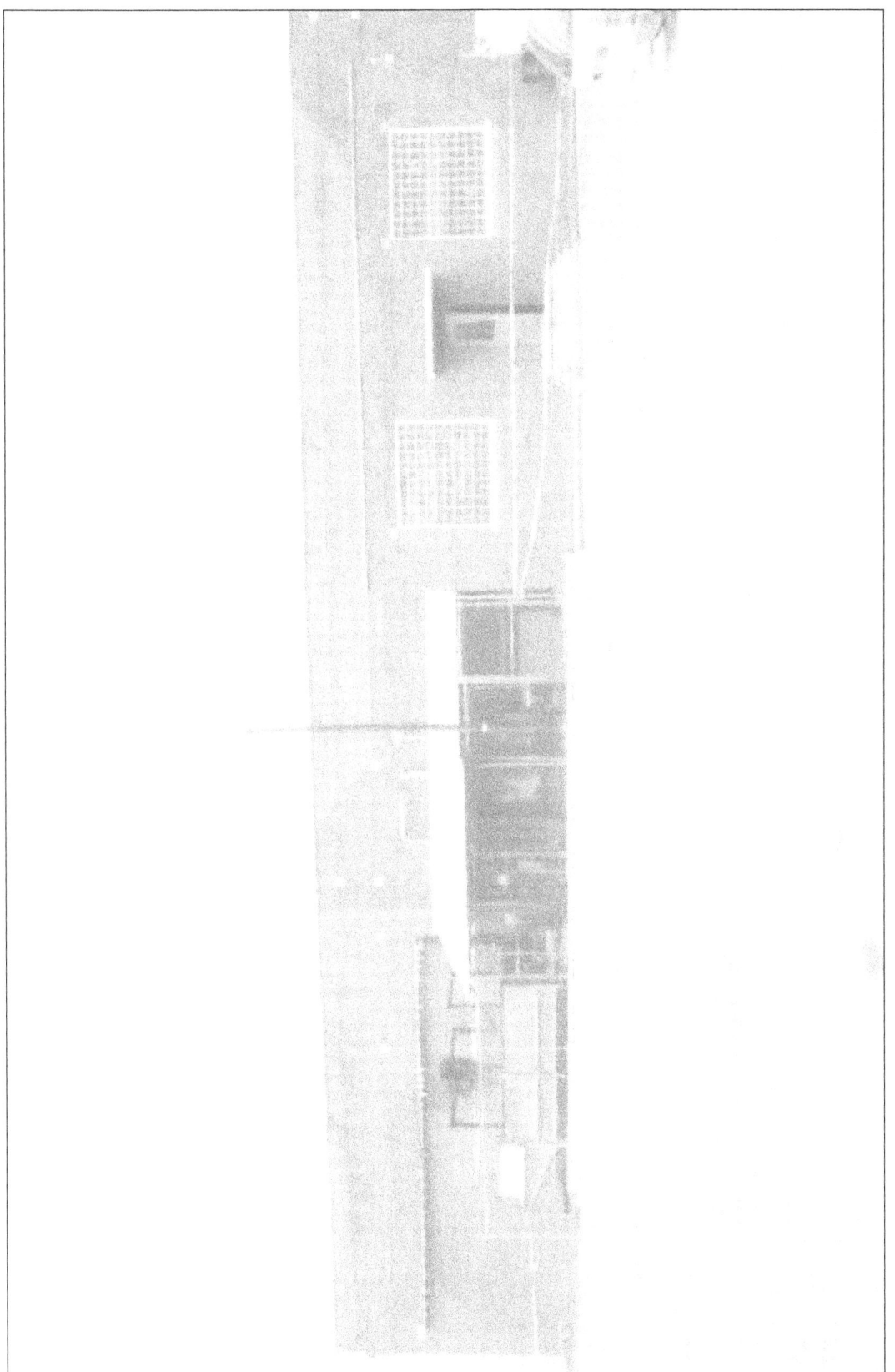

Photo by Jack Yates

The front of the main building. The door just right of center is to the front office area.

Appendix C (continued)

Photo by Jack Yates

A view from an aerial ladder from the east looking southwest. Some roof damage is visible at the extreme right.

Appendix C (continued)

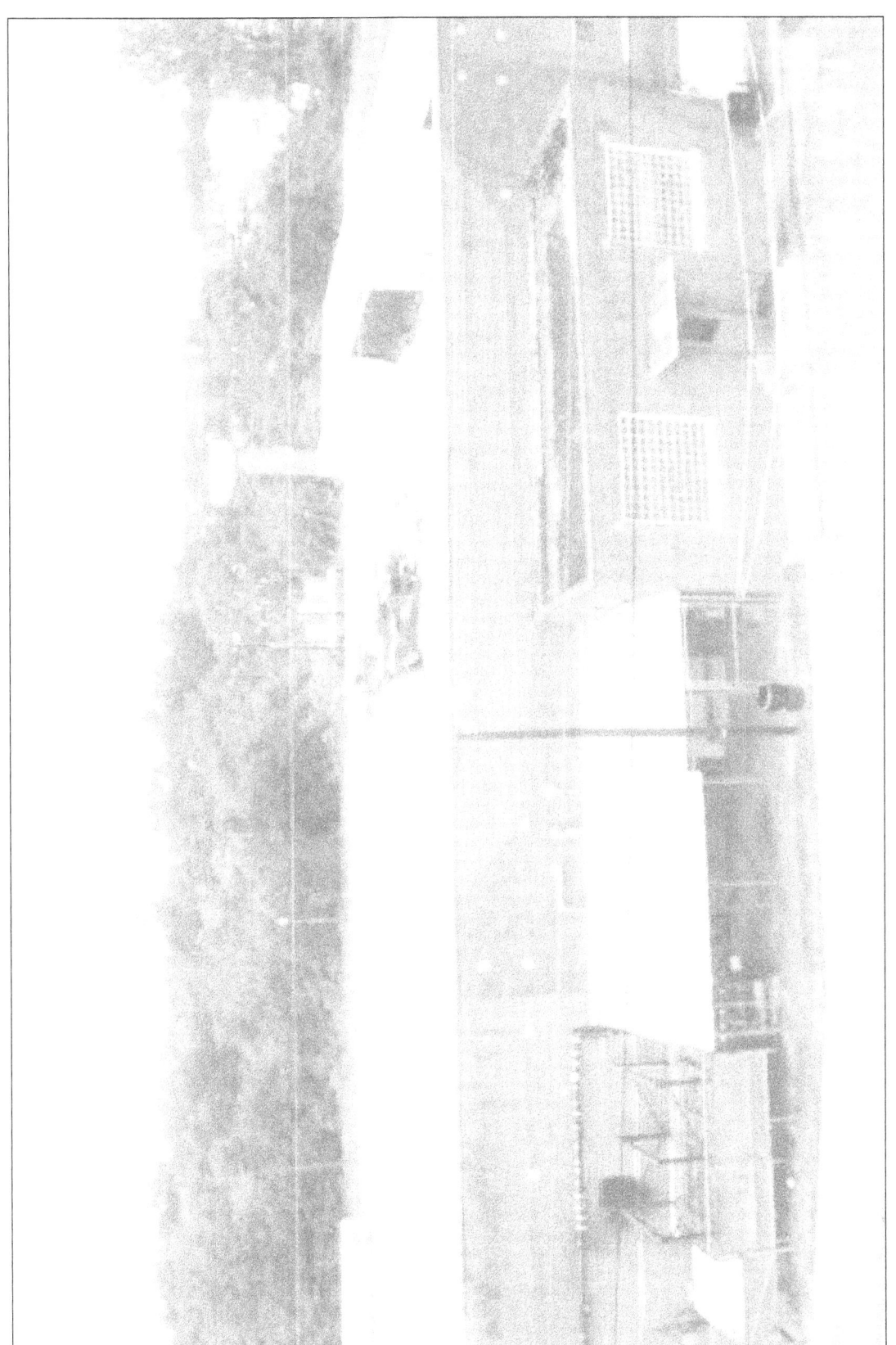

Another view from the aerial ladder from the east looking west, shows the portion of the roof collapsed over the area of origin.

Appendix C (continued)

Another view from the aerial ladder from the east looking northwest, shows the remainder of the building area.

Photo by Jack Yates

Appendix C (continued)

Photo by Jack Yates

A view from the aerial ladder from the west side of the building looking to the southeast. The break room roof is in the forefront.

Appendix C (continued)

Photo by Jack Yates

A view from the aerial ladder from the west side looking to the northeast. The roof collapse can be seen on the right, refrigeration compressor equipment is seen on the ground in the lower left.

Appendix C (continued)

Photo by Jack Yates

The trash and loading dock area. It was into this compactor area that several people tried to escape only to find the door to the outside locked. Upon coming back out of this area, they went through the large doors at the right which ultimately led to the large walk-in cooler where the largest fatality count was found. View is from the east toward the west southwest.

Appendix C (continued)

Photo by Jack Yates

View showing the trim roof as seen from the west center looking to the southeast. This view is looking toward the door to the cooler where the highest fatality count was located.

Appendix C (continued)

Photo by Jack Yates

Inside the marinating and cutting room; no direct flame impingement actually entered into this area. View is from the east to the west.

Appendix C (continued)

Photo by Jack Yates

The cooler door, which was at the south end of the building, where the highest fatality count was found.

Appendix C (continued)

Photo by Jack Yates

The west wall in the west portion of the cooler. Numerous handprints were found where people were trying to find escape routes.

Appendix C (continued)

Photo by Jack Yates

The trim room area showing the north end of the room and also a door leading to the break room area.

Appendix C (continued)

The east half of the break room, view taken from the southwest toward the northeast.

Appendix C (continued)

Photo by Jack Yates

A close-up view of the north door to the break room. Note the footprints and padlock in-place where the door was kicked; it was eventually opened for the escape of some personnel.

Appendix C (continued)

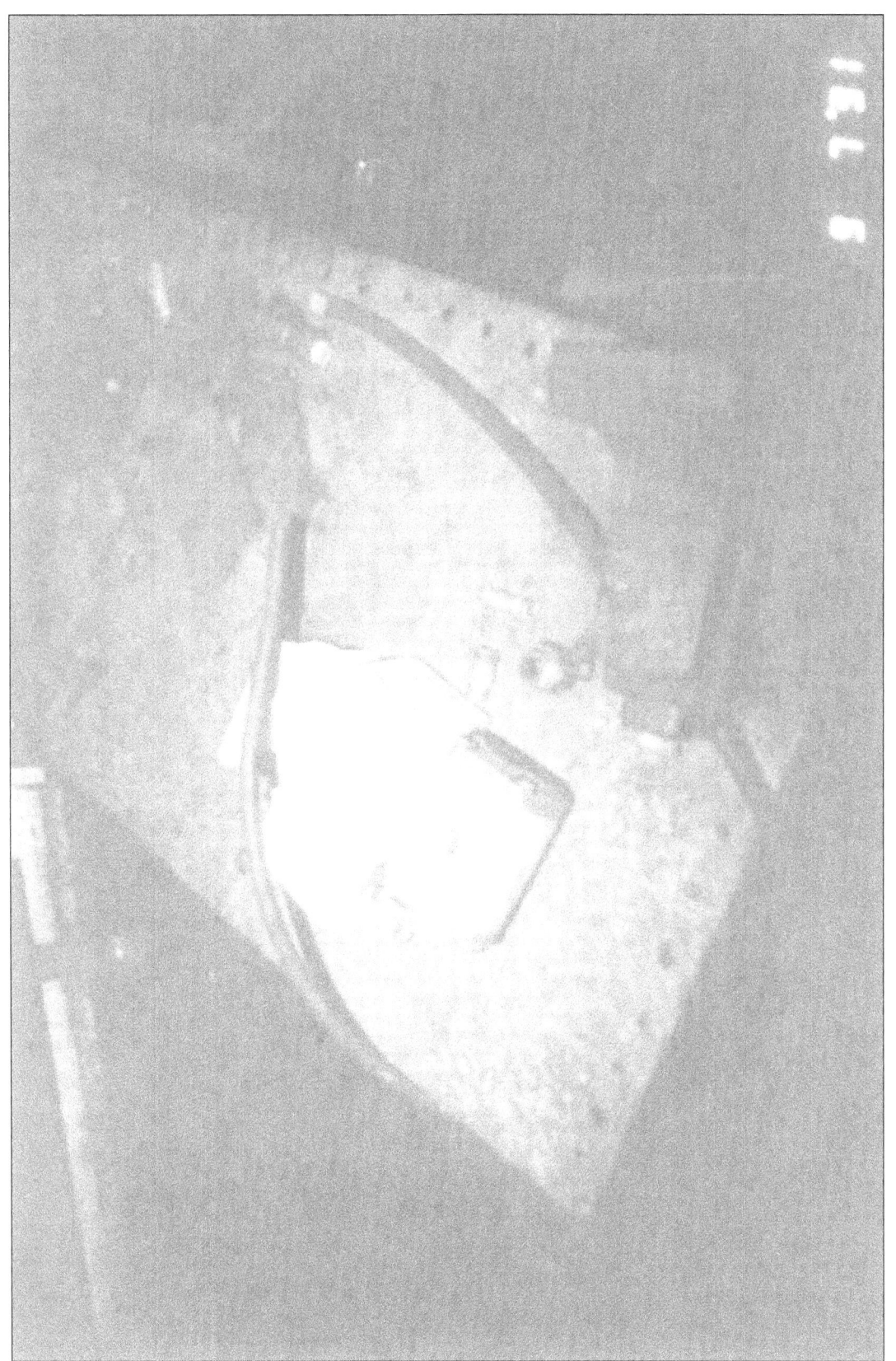

Photo by Jack Yates

The remains of the hydraulic line that had been cut by the maintenance worker for the repair process that took place just preceding the fire.

Appendix C (continued)

Photo by Jack Yates

The west side of the cooking vat. View taken from the north to the south.

Appendix C (continued)

Photo by Jack Yates

The area of origin viewed from the southeast to the northwest.

Appendix C (continued)

Damage to the steel girders and roof supports over the area of origin.

Photo by Jack Yates

APPENDIX D

Tyson Foods Safety Policy, Monthly Fire Inspections Checklist and Other Fire Safety Program Materials

TYSON FOODS INC.

SAFETY POLICY

INCIPIENT FIRE FORCE

Description: corporate Safety Policy relative to a Fire Force and Fire Protection within the Tyson organization.

Scope: This policy covers ALL Tyson facilities.

Individual Plant Requirements:

1. An Incipient Fire Force will be established at all Tyson facilities.

2. Emergency Action and Fire Prevention Plans will be prepared at each facility and copies submitted to Loss Control.

3. Exit drills will be conducted at least semi-annually or when Evacuation plans are revised.

4. Alarm systems will be installed per Section 1910.165 of OSHA.

5. The Fire Force will conduct quarterly training sessions.

6. Records will be kept as to training sessions, alarm tests, sprinkler tests, and fire hazard inventories.

7. All fires will be reported to Loss Control as soon as practicable and a follow-up written report made.

8. Fire extinguishers will be inspected monthly and records kept.

9. Each facility will obtain and install an adequate number of fire extinguishers and hose stations as required.

10. Monthly fire inspections will be conducted and copies of those inspections will be forwarded to loss Control not later than the 30th of each month.

Approved: _____ Date: 10-5-90

35

Appendix D (continued)

INTRODUCTION

As Tyson Foods *continues to expand* and *change, fire protection becomes more complex and difficult. New processes and products bring new fire hazards. Processing equipment and* facilities *have become even larger and more* expensive. Their *loss has a greater* impact *on production and the bottom line. Greater values* are *concentrated* in *single buildings.* Products *are stored higher* and *higher in warehouses. More and more personnel are concentrated and* exposed to greater *hazards.*

Fire detection and prevention equipment is hard pressed to *keep pace with the new hazards. As a result, the risk of very large losses is increased-- losses which can threaten the* entire *plant or event the entire business organization. Maintaining* these risks within reasonable *bounds is a major challenge to management.*

Good fire protection doesn't just happen, it *is the result of* corporate policies *and related* fire *prevention programs. Good organization,* with *responsibilities clearly assigned and specific duties* spelled out, *will result in implementation of* effective *programs.*

The two primary ways *to manage fire risks are to prevent fires and to limit or control their* size.

An effective *program* receives its *driving force and continuing motivation from top management, but strong interest extending down through various levels of management* and *supervision to the individual employee is needed for the program to succeed.*

The objectives of a satisfactory fire prevention and control program *can be stated very* simply:

1.   *To plan and construct low hazard buildings, processes, and equipment.*

2.   *To provide adequate fire Control and Suppression equipment where needed.*

3.   To *educate and train employees in loss* prevention *and* proper action *in emergencies.*

In *planning new facilities decisions made* during *the planning stages largely determine the degree of fire risk the facility will present* after *it is built. The important considerations are in the following* areas:

1.   *Safety to life*

2.   *Protection of property*

3.   *Continuity of operations*

Appendix D (continued)

CUTTING & WELDING
P E R M I T

Applies Only to Area Specified Below

Date_____

Building_____   Floor_____

Nature of the job_____

The above location has been examined. The precautions checked below have been taken to prevent fire. Permission is granted for this work

Permit expires:_____

Date_____   Time_____

Signed_____

Fire Safely Supervisor

Time started _____   Time finished _____

FINAL CHECKUP

Work area and all adjacent areas to which sparks and heal might have spread (such as floors above and below and on opposite side of walls) were Inspected for at least 30 minutes after the work was completed and were found fire safe.

Signed_____

After signing, return permit to person who issued it.

PRECAUTIONS

The Department supervisor or his appointee should inspect the proposed work area and check precautions taken to prevent fire.

☐ Sprinklers in service.
☐ Cutting and weiding equipment in good repair.

PRECAUTIONS WITHIN 35 FEET OF WORK

☐ Floors swept clean of combust bles.
☐ Combustible floors wet down, covered with dark sand or metal or fireproof sheets.
☐ No combustible materials or flammable liquids.
☐ Combustibles and flammable liquids protected with fireproof tarpaulins or metal shields.
☐ All wall and floor openings covered.
☐ Fireproof tarpaulins suspended beneath work to collect sparks.

WORK ON WALLS OR CEILINGS

☐ Construction noncombustible and without combustible covering or Insulation.
☐ Combustibles moved away from opposite side.

WORK ON ENCLOSED EQUIPMENT

Tanks, containers. ducts. dust collectors, etc.)

☐ Equipment cleaned of all combust bles.
☐ Containers purged of flammable vapors.
☐ Inlets 6 outlets locked out &plugged.

FIRE WATCH

☐ To be provided during and for 30 minutes after operation. recheck after 2 hours.
☐ Supplied wlth extinguishers or small hose.
☐ Trained In use of equipment and In sounding alarm.

Signed _____

(Tyson) 16147

Appendix D (continued)

```
                       TYSON FOODS, INC.

              MONTHLY FIRE INSPECTION CHECKLIST


   Facility                                    Date


1.  Fire Extinguishers

        a.   Was each Unit examined?

        b.   Were all Refills completed?

        C.   Were Units easily Accessible?

        d.   Condition of Units:



2.  Smoking Regulations

        a.   List "Smoking" Areas


        b.   Non - Smoking Areas Posted?

        C.   Regulations Enforced?

3.  Volatile and Combustible Materials

        a.   Were these Materials Needed where found?

        b.   Are Materials Safely Stored and Handled?

        C.   Are Safety Containers used and in good condition?

        d.   Any stored under Stairwells?

        e.   Any Excessive Amounts?
```

Appendix D (continued)

4. Fire Drills

    a. Date Held._____

    b. Was Drill Expected?_____

    c. Number of Persons in Drill._____

    d. Was signal clear to all persons?_____

COMMENTS:_____

_____

5. Hose Stations

    a. Was a Hose attached to each outlet?_____

    b. Was a Nozzle attached to each Hose?_____

    c. Is Hose Properly Racked?_____

    d. What Condition are the hoses in?_____

    e. Date Tested._____

6. Sprinkler Systems

    a. Valves Open?_____

    b. Stand Pipes Inspected?_____

    c. Sprinkler Heads Un-obstructed?_____

    d. Sprinkler Heads Painted?_____

    e. Sprinkler Heads or Piping Corroded?_____

    f. Sprinkler Heads loaded with Dirt?_____

    g. Sprinklers obstructed by New Partitions?_____

    h. New Section requiring Sprinklers?_____

    i. Flow Test Conducted?_____

      _____   _____

        water pressure   pressure w/ drain valve open

Appendix D (continued)

7. City Water:

    a. Is City Water in Commission? _____

    b. Gage Pressure. _____

8. Steam Piping

    a. Are all pipes and coils one inch clear
        of wood work and supported safely? _____

    b. Ducting of Exhaust in safe condition? _____

9. Wiring and Electrical Equipment:

    a. Are all panel boards, switch and fuse
        cabinets clean? _____

    b. Are all outlet box covers in place? _____

    c. Are all fuse and switch box covers
        in place? _____

    d. Is there any temporary wiring? _____

        If so comment on Location: _____

10. Housekeeping

    a. List Locations where housekeeping was not satisfactory:

        _____

        _____

        _____

    b. Will these be cleaned up? _____

11. Detection Systems

    a. Heat Detectors in operable condition? _____

    b. Smoke Alarms in operable condition? _____

Appendix D (continued)

12. Manual Fire Alarms

    a. Are stations un-obstructed?_____

    b. Are stations operational?_____

13. Vent Hood Systems

    a. Semi-Annual Inspections completed?_____

    b. Are they clear of grease accumulation?_____

14. Exits

    a. Are there an ample number?   _____

    b. Continuously lighted and/or visible?   _____

    c. Are doors opened easily?   _____

    d. Are doors unlocked?   _____

    e. Are doors un-obstructed?   _____

    f. Has emergency lighting been tested?   _____

    g- Are there at least two Remote Exits?   _____

15. Building Exterior

    a. Are stated Fire Lanes un-obstructed?   _____

    b. Are Fire Hydrants easily accessible?   _____

    c. Are sprinkler valves in open position? _____

    d. Are Fire Exits blocked?   _____

16. Inspections:

    a. Are end-of-work-day Inspections being done? _____

17. Fire Brigade Training

    a. Are monthly training sessions being conducted? _____

Appendix D (continued)

18. Construction Areas

    a. Check for Fire Hazards.   _____

    b. Check for Block Exits of Fire Lanes. _____

Comments: _____

_____

_____

    I HEREBY, CERTIFY THAT I HAVE INSPECTED THE ABOVE LISTED ITEMS AS SHOWN BY THE COMMENTS THERON.

    (*TO BE COMPLETED BY THE 20TH OF THE EACH MONTH)

_____   _____
    Fire Brigade Chief   Date

_____   _____
    Facility Manager   Date

Appendix D (continued)

Tyson Fire Safety
(Fry Department)

The following is a list of controls in place at virtually all locations with fry operations. Hydraulic line interlock valve installation should be completed company-wide within 2 weeks. Supervisor training on fire extinguisher operation is being updated -

1. Fryer Suppression System -
    200 lb. Co2 Automatic Extinguishing System, 3 minute discharge.

2. Fryer Gas-line Interlock- Electronically shuts off natural gas supply in the event of a fire.

3. Fryer Hydraulic line interlock -
    Shuts off hydraulic fluid flow in the event of a ruptured line.
    (These are currently being installed).

4. Quarterly Suppression System Maintenance.

5. Available, strategically located portable fire extinguishers.

6. Supervisors Trained to use Extinguishers.

7. Automatic Fire Alarm Systems - * not installed in all locations.

8. Fire Retardant Room Construction.

9. Exits located for quick access.

10. Exit Instructions Communicated and Posted.

11. Exits maintained clear, unlocked and adequately marked.

12. Plant Fire Force Team (Fire Brigade) -on site- Regular Training and monthly meetings.

13. Emergency Response Equipment (Respirators, Chemical Suits, etc.) on site.

14. Written "Crisis Management" Plan - on site.

15. Routine Inspections - In house.

16. Local Fire Department Inspection/Consultation.

Appendix D (continued)

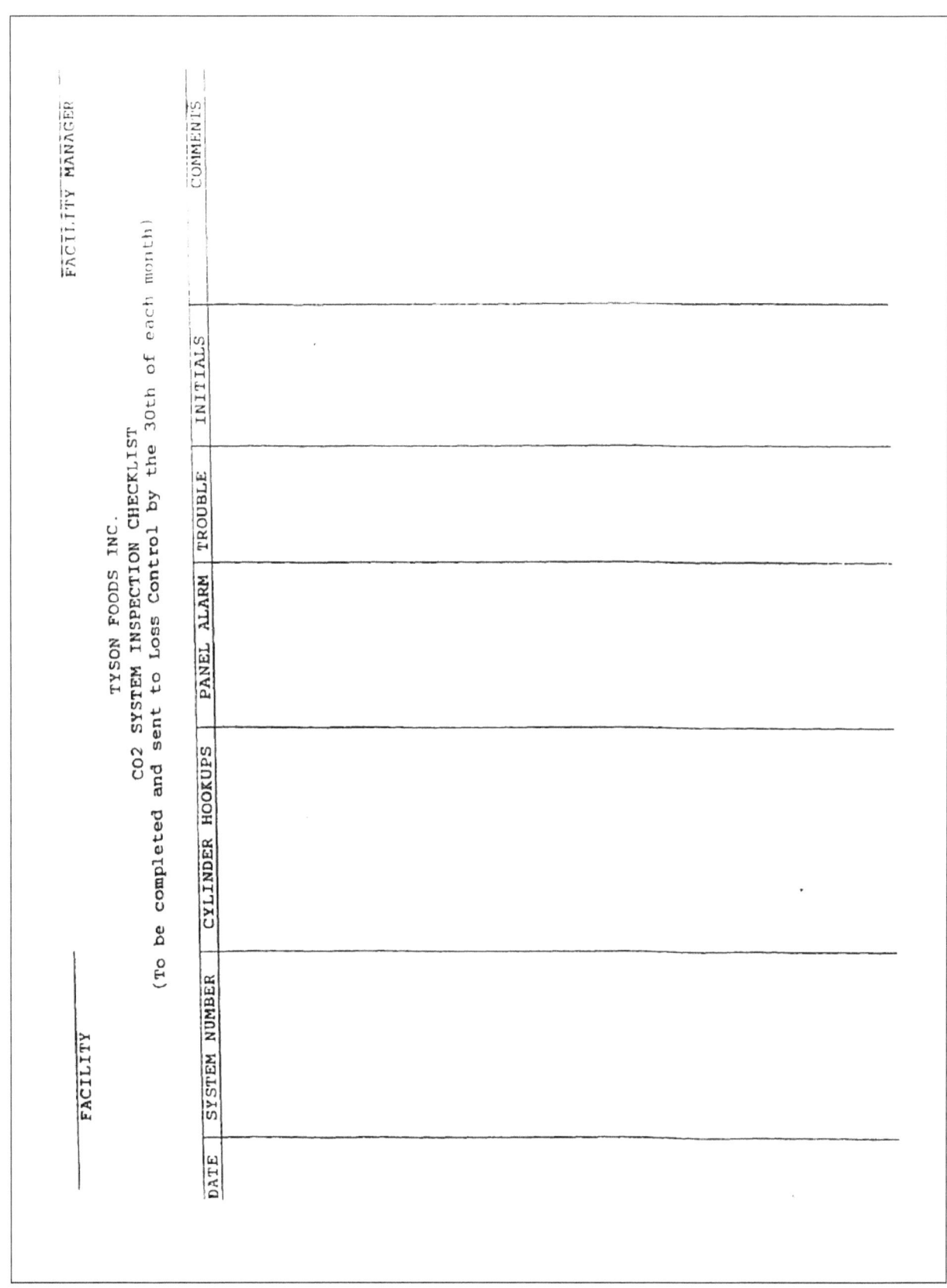

FACILITY _____   FACILITY MANAGER _____

TYSON FOODS INC.
CO2 SYSTEM INSPECTION CHECKLIST
(To be completed and sent to Loss Control by the 30th of each month)

DATE	SYSTEM NUMBER	CYLINDER HOOKUPS	PANEL ALARM	TROUBLE	INITIALS	COMMENTS